Edilson Ramos de Lima

Food Security with Productive Inclusion in the Semi-arid Region of Alagoas

Edilson Ramos de Lima

Food Security with Productive Inclusion in the Semi-arid Region of Alagoas

ScienciaScripts

Imprint

Any brand names and product names mentioned in this book are subject to trademark, brand or patent protection and are trademarks or registered trademarks of their respective holders. The use of brand names, product names, common names, trade names, product descriptions etc. even without a particular marking in this work is in no way to be construed to mean that such names may be regarded as unrestricted in respect of trademark and brand protection legislation and could thus be used by anyone.

Cover image: www.ingimage.com

This book is a translation from the original published under ISBN 978-613-9-64630-2.

Publisher:
Sciencia Scripts
is a trademark of
Dodo Books Indian Ocean Ltd. and OmniScriptum S.R.L publishing group

120 High Road, East Finchley, London, N2 9ED, United Kingdom
Str. Armeneasca 28/1, office 1, Chisinau MD-2012, Republic of Moldova, Europe
Printed at: see last page
ISBN: 978-620-7-78179-9

TABLE OF CONTENTS

CHAPTER 1

INTRODUCTION

"Nutritional and Productive Food Security in Rural Communities of the Ipanema Territory of Alagoas" was an initiative of the Consortium for the Development of the Ipanema Region - CONDRI - in conjunction with the consortium municipalities of the Ipanema Region of Alagoas, Territorial Collectives, Community Associations, Quilombola Communities and Civil Society Organisations working in the territory. It is a dialogue and partnership with the Secretariat for Food and Nutritional Security and the Ministry of Social Development and Fight against Hunger, with the aim of contributing to the implementation of structuring actions to make the **Brazil Without Poverty Programme** effective in the Sertão de Alagoas.

The development of integrated production systems in rural communities is an emancipatory proposal to promote the human right to adequate food for families living in extreme poverty and social exclusion. These are organisational, training and production actions aimed at structuring agri-food and associative activities to strengthen family production units.

CONDRI structured these activities with six (6,000) families and thirty (30,000) people from the semi-arid region, with the aim of promoting self-sufficiency and productive autonomy for these families. With the implementation of the project, it helped to eliminate food insecurity among families and promoted social and productive inclusion, insofar as integrated production systems were structured, native seeds and products were exchanged from surplus food consumption and experiences were exchanged between family units.

The project was structured from the perspective of also promoting education for citizenship and valuing the cultural diversity of the families; guaranteeing the management and conservation of the genetic variety of seeds, plants and animals native to the caatinga; favouring food and nutritional adequacy and developing associative and solidarity-based forms of production and coexistence, capable of generating a fabric of social relations based on equity, cooperation and solidarity.

According to IBGE/2010 data, in Brazil there are still 16.267 million people (8.5% of the population) living on a monthly per capita income of up to R$70.00, despite the results achieved by public policies over the last 10 years[1] . This is the extreme poverty line defined by the Federal Government, which underpins the *Brazil Without Poverty Programme.* The largest concentration of people living in extreme poverty is in the Northeast, which is almost 60 per cent (9.61 million people) of the country's poor, of whom 56.4 per cent are in rural areas.

For the same period, according to recent studies by the Institute for Applied Economic Research (IPEA), the number of people living in absolute poverty is still alarming in the state of Alagoas. Alagoas leads the ranking of extreme poverty in the country, with 32.3 per cent, followed by Maranhão (27.2 per cent) and Piauí (26.1 per cent). Thus, according to the criterion of the proportion of poor people, Alagoas has moved from the seventh position it was in the 1980s to first place in the Northeast Region.

The picture of poverty in Alagoas worsens when we look at the situation of families living in the semi-arid region of the state. The scenario of contradictions and social and economic injustices characteristic of the historical occupation of space has transformed this region into an environment marked by intense concentration of land and income, poverty and violence.

The precarious living conditions suffered by a large number of families in the semi-arid region are exacerbated during periods of prolonged drought, due to difficulties in accessing water in quantity and quality for human consumption and production.

---

[1] According to data from the Institute for Applied Economic Research (IPEA), in the last 10 years 12.8 million people have been lifted out of absolute poverty in Brazil.

The situation of food insecurity, the lack of minimum conditions to produce, the lack of resources to buy quality food, makes families dependent on emergency government programmes. These are investments made in order to minimise the situation of hunger and the structural exclusion of families without, in isolation, promoting any kind of emancipatory action. Productive inclusion and food security become basic and essential conditions for the socio-economic sustainability of families in the semi-arid region of Alagoas.

As an alternative to tackling the issues mentioned above, this project has developed a programme to organise and train families to structure integrated production systems, with a view to self-sufficiency in food and nutrition, productive autonomy and qualified access to the federal government's public policies.

In this sense, the training and production processes have been structured to provide families with the necessary conditions to produce their own food and commercialise surpluses, thus increasing family income. The production systems are made up of diversification and proper management of plant production; the installation of small animal farms; the formation of gardens using Creole seeds and seedlings of native species; and access to water for production. These systems are appropriate to the reality of the semi-arid region and are based on the principles and practices of management and conservation of the caatinga's agrobiodiversity.

Food and nutritional security will also be the result of training families in production activities, the proper processing of food, the diversification of food components and the management of water resources for human and animal consumption. To this end, young mobilisation/training agents will be trained to act as facilitators of the training process for family production groups.

The combination of knowledge and experience, permanent dialogue, the conscious adherence and commitment of families, and the construction of participatory management are all essential methodological aspects for carrying out the constructive processes of this proposed action.

The aim of the partnership with the Ministry of Social Development was to overcome the challenge of promoting conditions for families in the semi-arid region to overcome poverty and extreme poverty in the northeastern state with the highest rates of social and productive exclusion, and to transform the results into lasting actions that can be disseminated to other territories, contributing to the consolidation of ***Brazil Without Poverty.***

CHAPTER 2

PROPOSER

CONDRI - Consórcio para o Desenvolvimento da Região do Ipanema do Estado de Alagoas (Consortium for the Development of the Ipanema Region of the State of Alagoas) is a non-profit public association set up in 2006, with its administrative headquarters at Rua Castelo Branco 84, Santana do Ipanema/AL. It was created as a result of coordination between civil society organisations and the municipalities of the mid-Sertão region of Alagoas. In its formation phase, the encouragement and support of Cáritas Brasileira Regional NE II and the Polis Institute of São Paulo/SP and organised civil society entities, aimed at greater interaction between the social and institutional players whose purpose is to enable joint actions to carry out shared purchases and/or contract common services, through a Central Purchasing Centre, using a bidding process; rationalise investments in order to achieve large-scale savings; carry out planning, regulation and inspection activities for purchases made by the consortium municipalities; promote technical training activities for staff in charge of managing and executing shared purchases by the consortium entities; respond to requests from consortium entities, carry out shared tenders which result in contracts signed by consortium entities or bodies of their indirect administration and oversee the execution of the contract (art. 112, § I° , of Law no. 8.666/1993); carry out planning, regulation and inspection activities for public solid waste services in the territory of the consortium municipalities; provide a public solid waste service or an activity that is part of a public solid waste service through programme contracts it signs with the interested parties; contract with a waiver of bidding, under the terms of item XXVII of the caput of art. 24 of Law no. 8.666, of 21 June 1993, associations or cooperatives formed exclusively by low-income individuals recognised as recyclable material collectors to provide collection, processing and marketing services for recyclable or reusable solid urban waste in areas with a selective waste collection system; to plan, regulate and supervise the management of construction waste and bulky waste, as well as, under the terms authorised by a resolution of the General Meeting, other waste for which the generator is responsible, to set up and operate; a network of delivery points for small quantities of construction waste and bulky waste; facilities and equipment for transhipment and sorting, recycling and storage of construction waste and bulky waste; setting up and operating collection services, facilities and equipment for the storage, treatment and final disposal of health service waste, under the terms of the contract with consortium entities and without prejudice to the responsibility of generators and transporters, observing the provisions of current federal legislation; promoting social mobilisation activities; promoting the rational use of natural resources and environmental protection; promoting technical training activities for the staff in charge of managing the public solid waste services of the consortium members; being contracted to provide technical assistance services to the bodies or entities of the consortium members on issues of direct or indirect interest to solid waste (art. 2 , § I). 2º , § Iº , III, of Law no. 11.107/2005); to a non-consortium municipality or private entity, provided that the priorities of the consortium members and the Regionalisation Master Plan (PDR) of the state of Alagoas are not prejudiced; To plan, schedule and execute programmes, projects, actions, activities and services in the health area, in accordance with the objectives set out in this clause; to share financial, technological and people management resources, and the common use of equipment, maintenance services, information technology, bidding procedures, service provider units, management instruments, especially care programming and the consortium's management plan, among others, in compliance with regionalisation rules; represent all its members in matters of common interest and of an environmental nature for fisheries, before any public or private, national or international entities; plan, adopt and execute plans, programmes and projects designed to promote and accelerate the sustainable development of fisheries and environmental conservation; to promote programmes and/or measures aimed at the recovery, conservation and preservation of the environment, with special attention to soils; mountains; plains, lagoons and lakes; rivers and streams with a view to improving fish farming; to promote the integration of actions, programmes and projects developed by government bodies and private companies, whether or not in consortium, aimed at the recovery, conservation and preservation of the environment with a view to improving fish farming; managing financial and technological resources with public bodies, financial institutions and the private sector for the sustained development of the region; making joint efforts through actions aimed at the integrated development of tourism, history and culture in the region; liaising with national and foreign public and private bodies with a view to planning and obtaining resources for investment in tourism projects, works or services; promoting tourism in the region; developing and promoting tourism sustainability in the region; seeking solutions for the social and economic development of tourism; preserving the archaeological memory, promoting socio-economic, historical, cultural, tourist, landscape and ecological development along the stretch of the railway; awakening the municipalities involved to tourist activity, through the history, culture and products of each municipality, helping them to discover their potential; seek funding for the consortium both through transfers from the federal and state governments, as well as through apportionment between the municipalities involved; sign covenants, contracts and agreements of any kind, receive aid, contributions and social and economic subsidies from other entities and government bodies (article 2º, § 1º , I of Law 11. 107/05), at all levels.107/05), at all levels, as well as from private individuals; to be contracted by the direct and indirect administration of the consortium federation entities, with no need for a bidding process, for the provision of services, even enjoying the increase in values provided for in the Bidding Law, for cases of waiver; to promote expropriations and institute services under the terms of the declaration of public utility or need, or social interest, made by the Public Power; to contract credit operations, always subject to the limits and conditions established by the Federal Senate, in accordance with the provisions of art. 52, item VII; and promote claims, studies and proposals with federal and state bodies of common interest to the members; promote sustainable rural family development in the municipalities located in the area of operation of this consortium, as well as other productive initiatives; carry out collective actions aimed at raising funds and expanding federal and state programmes in the consortium municipalities, also promoting coordination with governmental, non-governmental and international entities; planning ways to promote sustainable family development, creating joint mechanisms for consultation, study, execution, inspection and control of activities that interfere in their area, especially with regard to urban development and land use control.

In its eight years of existence, CONDRI has been working to strengthen family farming in the 17 consortium municipalities, developing structuring actions for the population's access to water for human consumption, developing and disseminating social technologies, qualifying family farmers' access to public policies, organising and planning territorial actions based on the adherence and commitment of municipal public managers and the support of civil society.

2.1 Consortium municipalities:

CONSORTIUM MUNICIPALITIES
Cacimbinhas
Canapi
Sheep
Dois Riachos
Inhapi
Major Izidoro
Marvellous
Monteirópolis
Olivença
Olho D'água das Flores
White Gold
Palestine
Sugar Loaf
Poço das Trincheiras
Santana do Ipanema
São José da Tapera
Senator Rui Palmeira

CHAPTER 3

OBJECTIVES

3.1 GENERAL

Families in rural communities in the hinterland of Alagoas, with access to public policies to promote social inclusion and income generation, are overcoming poverty and extreme poverty and achieving food, nutritional and water security and productive autonomy.

3.2 SPECIFIC

- Families with access to and control of water in quantity and quality for human consumption and production using rainwater harvesting equipment and the sustainable management of existing water resources in the municipalities.

- Family production units with diversified production based on access to water, small animal farms, native seed banks and productive backyards, ensuring adequate food and nutrition, productive autonomy, conservation of the caatinga and supplementing family income.

- The exchange of experiences between family production units on production processes suited to the semi-arid region, boosting cooperation ties between families and encouraging socialisation, participation, the construction of collective identities and associative and solidarity forms of production.

- Communities organised in associations, rights defence networks, councils, expanding ties and collective solidarity practices and acting in the proposition and social control of public policies.

CHAPTER 4

AUDIENCE AND SCOPE

- **6,000 families (30,000 people)** from family farming communities, quilombola communities, settled families and fishermen living in poverty and extreme poverty, without access to production and in a situation of social exclusion, according to the indicators adopted by the Federal Government's Brazil Without Poverty Programme.

- **17 municipalities in the Ipanema Region,** consortium members of CONDRI and belonging to the Citizenship Territories of Médio Sertão, Alto Sertão and Bacia Leiteira.

TABLE 01 - POPULATION OF THE MUNICIPALITIES ASSISTED BY THE PROJECT

POLO	MUNICIPALITIES	PROJECT AUDIENCE	
		N° Families	No. of people
01	INHAPI	150	750
	CANAPI	550	2.750
	WHITE GOLD	500	2.500
	WONDER	300	1.500
	TRENCH WELL	900	4.500
02	SANTANA DO IPANEMA	700	3.500
	OLIVENCE	240	1.200
	JARAMATAIA	60	300
	OLHO DAGUA DAS FLORES	170	850
	TWO RIVERS	210	1.050
	CACIMBINHAS	150	750
	MAJOR IZIDORO	90	450
	MONTEIRÓPOLIS	100	500
03	MEATS	190	950
	SUGAR BREAD	450	2.250
	SENATOR RUI PALMEIRA	520	2.600
	PALESTINE	140	700
	SÃO JOSÉ DA TAPERA	580	2.900
	TOTAL AUDIENCE	**6.000**	**30.000**

CHAPTER 5

OPERATIONALISATION OF TARGETS

5.1 - PROGRAMME IMPLEMENTATION OF TARGETS

TARGET1
Implementation of integrated production systems suitable for self-sufficiency and food and production autonomy for 6,000 families in the Ipanema region of the state of Alagoas.

STAGE 1.1 - Implementation of the Training Process with Local Development Agents and communities.

a) Activity:	Training meetings with groups of local development agents and social educators.
b) Objectives:	• Presentation of the Development Agents by the Consortium Members. • To socialise the project's actions and strategies for the formation of appropriate integrated production systems; • Draw up the six-monthly operational plan of activities; • Define the municipalities and groups of beneficiary families per mobilising agent.

a) Activity:	Mobilisation and awareness-raising meetings with the communities participating in the project and the formation of the Community Steering Committee.
b) Objectives:	• Mobilisation/sensitisation of families and communities to join and participate in the project; • Training families on the project's actions for setting up integrated production systems; • Collectively identify production alternatives and trends in each locality; • Building an information bank on the productive, socio-economic, cultural and environmental conditions of each community.

a) Activity:	**Training workshops on Food and Nutrition Security and participatory processes.**
b) Programme content:	<ul><li>Community organisation and participatory processes;</li><li>Solidarity economy, solidarity associations and cooperatives;</li><li>The human right to adequate food and the enforceability of rights;</li><li>0 what is food and nutrition security and the Brazilian FNS legislation;</li><li>Family farming and FNS;</li><li>Natural food processing and storage system;</li><li>Making use of production leftovers in the nutritional composition of food;</li><li>Collective food production and proper management of systems;</li><li>Practical food recipes for quantity production.</li></ul>
b) Objectives:	- Train Mobilisation Agents and local partners to advise, accompany and monitor the implementation of the project.

a) Activity:	**Training workshops on the management and conservation of Creole seeds, medicinal plants and native species of the caatinga to restore vegetation cover and multiple use in the production system.**
b) Objectives:	- Levelling up of Mobilisation Agents on knowledge and practices of management and conservation of native seeds, medicinal plant beds and seedling nurseries of caatinga species.
c) Programme content:	<ul><li>Education for coexistence with the semi-arid region.</li><li>Techniques for the production, management and cultivation of medicinal, native, fruit and exotic plants;</li><li>Production techniques, management and selection of seedlings, cultural treatments and reforestation processes;</li><li>Enriched nutrition, production of home remedies and preventive alternative medicine.</li></ul>

a) Activity:	Training workshops on proper handling of small animals (sheep, pigs and poultry).
b) Objectives:	- Levelling up of Mobilisation Agents on knowledge and management practices for goats and sheep in the semi-arid region.
c) Programme content:	• The importance of feeding colostrum or first milk; • Treatment or cure of the navel; • Permanence of the cubs on the premises; • Feeding suckling chicks; • Weaning and the separation of offspring by sex; • Feeding weaned chicks; • Identifying the cubs • Decoma of the rams • Castration; • Health and hygiene;

a) Activity:	Organisation of SISMA training workshops.
b) Objectives:	- Integrated family production units practising techniques suited to the principles of coexistence with the semi-arid region.
c) Programme content:	• Sustainable production systems; • Production planning in integrated production systems in the semi-arid region; • Raising small animals, setting up simplified systems, native seed banks and access to and management of water for production.

a) Activity:	Holding training workshops on GAP A.
b) Objectives:	- Training beneficiary families from family production systems in the conservation and preservation of water resources in the communities.
c) Programme content:	• 0 combating drought and living in the semi-arid region; • 0 caatinga biome; • The caatinga and its potential; • Degradation and desertification processes in the caatinga: causes, consequences and impacts on the environment;

	• Available water resources in the semi-arid region; • Mapping existing water resources in the communities; • The collection and use of rainwater for human and animal consumption; • Techniques for conserving and preserving water resources in communities.

STAGE 1.2 - Implementation of the Training Process with beneficiaries of Integrated Production Systems in the Ipanema Region.

a) Activity:	Training workshops for bricklayer farmers in techniques for building social technologies in the semi-arid region.
b) Objectives:	- Train bricklayer farmers in social technology to capture rainwater for production
c) Programme content:	• Notions of water resource management; • Water for human consumption and production through rainwater harvesting and storage. • 0 what is social technology? Materials for building the technology; dimensions; construction stages. • Practical class on building social technologies.

a) Activity:	Training workshops on the organisation, management and conservation of seeds and native species of the caatinga.
b) Objectives:	- Valuing the culture and practice of producing and selecting native seeds for food security and productive autonomy.
c) Programme content:	• Identification and recording of the region's native seeds; • Techniques for the production, handling, cultivation and storage of creole seeds; • Formation of production beds and community seed banks. • Family production planning.

a) Activity:	Training workshops on health and reproductive management for small animals in the semi-arid region (sheep, poultry and pigs).
b) Objectives:	- Integrated family production units practising appropriate health and reproductive management techniques for sheep and pigs.
c) Programme content:	• Asepsis of facilities and equipment used in breeding centres; • Notions of physiological, pathogenic and parasitic disease control; • Use of medicinal plants in disease prevention; • Reproductive management techniques for sheep, poultry and pigs.

a) Activity:	Organisation of SISMA training workshops
b) Objectives:	- Integrated family production units practising techniques suited to the principles of coexistence with the semi-arid region.
c) Programme content:	• Sustainable production systems; • Production planning in integrated production systems in the semi-arid region; • Raising small animals, setting up simplified systems, native seed banks and access to and management of water for production.

a) Activity:	Training workshops on GAPA and coexistence with the semi-arid region.
b) Objectives:	- Training beneficiary families from family production systems in the conservation and preservation of water resources in the communities.
c) Programme content:	• 0 combating drought and living in the semi-arid region; • 0 caatinga biome; • The caatinga and its potential; • Degradation and desertification processes in the caatinga: causes, consequences and impacts on the environment; • Available water resources in the semi-arid region; • Mapping existing water resources in the

	communities; • The collection and use of rainwater for human and animal consumption; • Techniques for conserving and preserving water resources in communities.

Target02 - Implementation of integrated production systems in rural communities in the Ipanema region

a) Activity:	Construction of 24,000 litre cisterns.
b) Objectives:	- Installation of cisterns in family production units to develop the integrated production system.
c) Programming:	• Purchase of materials to install the cisterns; • Technical guidance for families and participation in the installation process; • Technical guidelines for handling and managing water from cisterns.

a) Activity:	Purchase of artisan kits for building social technologies.
b) Objectives:	- Installation of social technologies in family production units to develop the integrated production system.
c) Programming:	• Purchase of inputs for the installation of social technologies; • Technical guidance for bricklayer farmers on the proper use of tools for installing social technologies.

a) Activity:	Installation of tarpaulins with pavements for small-scale family production.
b) Objectives:	- Install dams to collect and store water for the production of family units.
c) Programming:	• Purchase of materials for the installation of tarpaulins; • Technical guidance for families and participation in the installation process; • Technical guidelines for the management of water in clay pits.

a) Activity:	Installation of rustic paddocks in family production units to develop an integrated

	production system and capture rainwater.
b) Objectives:	- Install rustic paddocks in family production units to develop the integrated production system
c) Programming:	• Purchase of materials for the installation of the paddocks; • Technical guidance for families and participation in the installation process; • Technical guidelines for the proper management of sheepfolds.

a) Activity:	**Installation of small ponds for family production.**
b) Objectives:	- Install dams to collect and store water for the production of family units.
c) Programming:	• Purchase of materials for the installation of the barraginhas; • Technical guidance for families and participation in the installation process; • Technical guidelines for the management of water in clay pits.

a) Activity:	**Installation of simplified irrigation kits in production units to develop the integrated system.**
b) Objectives:	- Install simplified irrigation systems in family production units to develop the integrated production system
c) Programming:	• Purchase of materials for the simplified irrigation system; • Purchase of vegetable and fruit seeds; • Acquisition of seedlings of medicinal plants from the caatinga; • Technical guidance for families and participation in the installation process; • Technical guidelines for the proper management of simplified systems.

a) Activity:	**Installation of shelters and purchase of animals, fodder and feed for family production systems with sheep, pigs and poultry.**
b) Objectives:	- Acquisition of animals for family production

	units and development of the integrated production system
c) Programming:	• Acquisition of animals to set up the breeding centres; • Technical guidance for families and participation in the process of setting up farms; • Technical guidelines for the proper management of farms.

a) Activity:	**Installation of swiddens to guarantee food support for families and animals with the purchase of seeds and storage and feed depots.**
b) Objectives:	- Installation of seed banks for family production units and purchase of seeds and silos for storage.
c) Programming:	• Acquisition of seeds for the installation of Creole seed banks; • Technical guidance for families and participation in the process of setting up and managing seed banks; • Acquisition of silos for seed storage and conservation; • Technical guidelines for the proper management of seed banks.

a) Activity:	**Purchase of work kits for Local Development Agents to provide technical assistance to production systems.**
b) Objectives:	- Encourage the provision of technical advice to family production units on the development of integrated production systems.
c) Programming:	• Purchase and distribution of travel kits for Development Agents; • Provision of technical advisory services to family production units; • Technical guidelines for the proper handling of veterinary kits.

TARGET 03 - Meetings to plan actions and evaluate the programme's results.

a) Activity:	Meetings to plan actions and evaluate the programme's results.
b) Objectives:	- Planning actions and defining strategies for carrying out activities and evaluating the results of actions.
c) Programming:	• Action planning meetings; • Preparation of six-monthly operational plans; • Procedural and results evaluation; • Meetings with partners, technical coordination and the implementation team.

GOAL 04 - Training the Technical Team to implement the programme.

a) Activity:	Hiring the project's technical assistance team and other permanent materials.
b) Objectives:	- Ensure the implementation of the programme's actions, liaise with movements and partners.
c) Programming:	• Definition of selection criteria and technical profile for the programme; • Publication and dissemination of the selection notice; • Composition of the selection board; • Selection and hiring of the technical team.

GOAL 05 - Socialising the training methodology and results with other organisations and social movements.

a) Activity:	Systematisation of the methodology applied to the project's training and production processes.
b) Objectives:	- To socialise the methodological procedures developed in the formation processes of collective family production systems.
c) Programming:	• Definition of instruments for recording experiences; • Creation of the database and information; • Documentary video production; • Production of educational booklets and bulletins; • Publication in periodicals.

CHAPTER 6

METHODOLOGY

The proposed methodology is based on participation and is structured into several stages, the guiding principles of which are: ***a) to stimulate purposeful dialogue and participatory management among the various players; b) to recognise and value the socio-cultural universe of the communities; c) to promote an educational environment that fosters local knowledge, the exchange of experiences, collective organisation and the strengthening of cultural identities; and d) to encourage solidarity and cooperation practices among the participants in the actions.***

In this sense, the development process is understood as the result of the collective construction of the various social actors, the improvement of associative capacity, the exercise of initiative and creativity. It is a process of social, cultural, political and economic empowerment that happens when social forces that were dispersed become convergent, take ownership of their problems and commit to solving them.

The training, organisational and production activities proposed in this project are based on four strategic axes:

1. Promote the management of water resources and ***access to water for consumption and family production.***

2. Developing productive alternatives to ***ensure food security and autonomy for families*** based on the principles and practices of living in the semi-arid region.

3. Promote and qualify ***families' access to public policies for social and productive inclusion and income guarantee.***

4. Developing participatory methodologies that stimulate *organisation, skills and opportunities, and solidarity and cooperation practices* among participants.

6.1 STAGES   OF THE TRAINING AND PRODUCTION PROCESS

a) Mobilising / coordinating families:

Local committees will be set up to monitor the actions and identify the families that will take part in the project, according to criteria defined collectively based on the poverty profile indicators adopted by the Brazil Without Poverty Programme. Once the groups of families participating in the project have been defined collectively, a workshop will be held to socialise and debate the proposed methodology and identify representatives for each of the project's hubs.

b) Training development agents:

The group of Development Agents is made up of **rural young people,** each of whom will accompany a group of no more than 40 beneficiaries from the communities participating in the project. These young people will be selected in the schools on the basis of their interest in taking part in the project, their interest in community organisation activities, their knowledge of the local situation and their interest in agricultural activities. The training workshops alternate theoretical and practical on-site training with the application of the knowledge acquired in the family production units. The aim of the Development Agents training is to enable young people to provide technical advice to the family production units set up by the project. The training sessions will cover popular education practices, associativism/cooperativism, food and nutritional security, agroecology and management techniques for integrated production systems that integrate sustainable and solidarity-based socio-economic activities. The mobilising agents who will accompany the productive nuclei will be appointed by the social movements in accordance with the beneficiary communities.

c) Holding mobilisation and awareness-raising meetings with the communities participating in the project:

These mobilisation and awareness-raising meetings will take place in one meeting per municipality. The meetings are scheduled for 048 hours and will be held with groups of 50 families. The methodology will be participatory and dialogical, valuing the families' experience and the exchange of knowledge and practices between the locals and the technical approach. It is a training tool that helps in the process of mobilising families to participate.

It also contributes to the collective identification of production alternatives and trends. It also contributes to the collective identification of production alternatives and trends; it provides an insight into the process of community organisation; it makes it possible to get to know the families' food consumption chain and their cultural values; it identifies levels of food security/insecurity, among other aspects of interest to the project. In this sense, the meetings are intended to fulfil a dual role by providing a primary source of up-to-date information and data on the communities (production, socio-economic, cultural and water conditions, etc.) and also to provide training by enabling the groups to become better acquainted with and participate in the project's activities, increasing their sense of belonging and commitment.

d) Implementation of the training programme for family production units:

Based on the information systematised in the Diagnosis, the programme content is made compatible and the teaching strategies for training family production groups are defined, identifying the main activities that will be developed in the family production units. The training programme is defined with the participation of development agents, productive groups and the local committee, divided into two modules: 1) a basic module, which should offer participants an environment for reflection on citizenship and organisation, the human right to adequate food, and planning the productive family unit; 2) a specific module, offering participants new knowledge and practices on water resource management and integrated productive systems suitable for living in the semi-arid region.

e) Installation of equipment to capture and store rainwater and family production systems.

Simultaneously with the implementation of the family training programme, the installation of rainwater harvesting and storage equipment and family production units began. Access to water for consumption and production aims to generate food for self-sufficiency and improve the quality of the families' diet, generate a surplus to guarantee the sustainability of the system and increase family income.

COMPARATIVE RELATIONSHIP BETWEEN THE PROJECT'S PRIORITY ACTIONS AND PROGRAMME 1133 / ACTION 4963 and PROGRAMME 1049 / ACTION 8948 - MDS.

PROGRAMME 1133 *Solidarity Economy in Development*		ACTION 4963 *Promoting Productive Inclusion*
PROGRAMME OBJECTIVE	PRIORITY PROJECT ACTIONS	MEASUREMENT INDICATORS
To promote the strengthening and dissemination of the solidarity economy through integrated policies aimed at generating work and income, social inclusion and the promotion of fair and solidarity-based development. *Action 4963* - To support and stimulate initiatives that present alternatives for generating work and income in areas made vulnerable by poverty and the lack of basic infrastructure, from the perspective of the solidarity economy.	Implementation of integrated production systems.	• Purchase of small animals for integrated production systems. • Installation of gardens for food production. • Travel kit for Local Development Agents.

PROGRAMME 1049 *Access to Food*		ACTION 8948 *Access to Water for the Production of Food for Self-Consumption*
PROGRAMME OBJECTIVE	PRIORITY PROJECT ACTIONS	MEASUREMENT INDICATORS

Objective	PRIORITY PROJECT ACTIONS	MEASUREMENT INDICATORS
To guarantee the population in a situation of food insecurity access to decent, regular and adequate food for nutrition and the maintenance of human health. *Action 8948* - Expand the conditions for capturing, storing and using water in production for self-consumption, based on the dissemination of information on the use of water experiences/technologies for sustainable land use and utilisation of water resources.	Training in Food and Nutrition Security and participatory processes.	- Local development agents trained and advising integrated production units.
	Training in the management and conservation of seeds, medicinal plants and native species of the caatinga.	- Local development agents trained and advising integrated production units.
	Training in proper handling of small animals such as sheep, pigs and poultry.	- Local development agents trained and advising integrated production units.
	SISMA training	- Local development agents trained and advising integrated production units.
	Training in GAPA and coexistence with the semi-arid region	- Local development agents trained and advising integrated production units.
	PRIORITY PROJECT ACTIONS	**MEASUREMENT INDICATORS**
	Training bricklayer farmers in techniques for building social technologies in the semi-arid region.	- Farmers and stonemasons trained in sustainable social technologies for storing rainwater from production cisterns.
	Training in the organisation, management and conservation of seeds and species native to the caatinga.	- Family farmers trained and managing seed banks.
	Training workshops on health and reproductive management for small animals in the semi-arid region (sheep, poultry and pigs).	- Family farmers trained and practising small animal management.
	SISMA training workshops.	- Family farmers capacitated and exercising the management of households simplified water systems.
	Training workshops on GAPA and coexistence with the semi-arid region.	- Family farmers trained and practising water management for food production.
	PRIORITY PROJECT ACTIONS	**MEASUREMENT INDICATORS**
	Aprisco cistern with a capacity of 24,000 litres	- 5,980 families
	Installation of tarpaulins with pavements for small-scale family production with 108,000 litre capacity	- 20 families
	Installation of rustic paddocks in family production units to develop the integrated	- 6,000 families

	production system and capture rainwater.	
	Installation of small reservoirs with a capacity of up to 500,000 litres for family production	-6 ,000 families
	Installation of simplified irrigation kits in family production units to develop the integrated system.	- 6,000 families
	Installation of seeds in productionandpurchasing seeds and storage.	- 6,000 families

Table 04 below shows the main operational procedures for carrying out the project.

SPECIFICATION	OPERATIONALISATION
Financial execution	The financial management of the project is the responsibility of CONDRI's financial and accounting team, under the coordination of the Executive Secretariat, and is carried out in compliance with the legislation in force for the execution of public agreements. In operational terms, the financial team works in synergy with the project's technical team, establishing a process of integration, continuity and meeting the procedural demands for carrying out the actions envisaged in the project.
Hiring the project's technical assistance team.	The technical team responsible for implementing the project's actions/activities will be hired through a public notice and selection process (CV analysis, proof of experience and interviews). When carrying out projects, CONDRI seeks to prioritise the hiring of technicians from the municipalities involved in the actions, when the technical requirements necessary for carrying out the actions are met.
Liaising with town halls	Consortium municipalities take part in regular meetings called by the Consortium. These are participatory meetings to define strategies for the development of the region; monitoring of project actions; rendering of accounts, among other routine issues. CONDRI's

	Executive Secretariat maintains a permanent programme of visits to the municipalities to monitor actions and strengthen the links between the consortium's municipal managers.
Selection and mobilisation of families participating in the project	Formation of Local Committees to monitor the actions and identify the families that should take part in the project, according to criteria defined collectively based on the MDS poverty profile indicators adopted in the Brazil Without Poverty Programme.
Purchase of materials for the cisterns; seeds, seedlings and animals.	The purchase of components for the installation of integrated production systems will be carried out in accordance with the legislation for tendering processes, with the standards and technical criteria for each component. The formation of small animal farms will prioritise goat, sheep and pig species adapted to the region.
Installation of integrated family production units, training processes and technical advice.	The installation of the integrated production units takes place at the same time as the training processes and technical advice to the units. The process will be participatory and dialogical, associating technical knowledge with practical action.

6.2 SUSTAINABILITY

The Family Production Units proposed in the project comprise an integrated and diversified production system, technically estimated to meet the following demands in a sustainable manner: a) guaranteeing the families' food security and autonomy in the production process; b) guaranteeing the reproduction of the family economic unit; and c) generating a surplus for commercialisation and income generation for the families.

Each Integrated Production System consists of rainwater harvesting and storage equipment, small animal husbandry, the formation of native seed banks and productive backyards.

Goat, sheep and pig species adapted to the region will be prioritised in the formation of the small animal farms. The importance of family breeding of these small animals has historically been a survival strategy for families in the semi-arid region. It therefore plays an important social and economic role in the region, as it is an option for improving life in the areas most affected by prolonged drought, acting as a favourable factor for keeping people in the countryside.

In addition to the natural and edaphoclimatic conditions favouring a balance between the feeding and handling habits of these native animals and the principles of coexistence with the semi-arid region, the regions in which the project operates are geared towards this type of economic activity. Social technological resources, available family labour, interest in the region in developing the activity, the existence of equipment installed with public funds to process the products[2] are some of the considerable factors that lead to their use, on new technical bases for living with the semi-arid region, in production models and solidarity management for the development of production chains and income generation.

The strengths underpinning the implementation of these integrated production systems could be summarised as follows: a) it is a typical family production activity in the region; b) it has favourable zootechnical characteristics in terms of adaptability and reproduction; c) it has a lower growth rate than the net increase in demand in the respective production areas[3] ; d) low costs associated with the ability to open up new markets; d) it offers meat, milk and dairy products rich in the nutrients needed to overcome the malnutrition levels in the communities.

6.3.1 COMPONENTS OF THE INTEGRATED PRODUCTION SYSTEM:

Family production units are structured on the basis of the components listed in Table 04 to 4.3, making up an integrated production system to guarantee the family producer's self-consumption, the reproduction of the production unit and the commercialisation of surplus production.

TABLE 04 - INTEGRATED PRODUCTION SYSTEM 01

INTEGRATED PRODUCTION UNIT	QUANT.	FAMILIES
Small farmyard* (up to 500,000 litres)	01	5,980 Families
Cistern* (up to 24,000 litres)	01	
Sheep animals	03	

---

[2] In the region, there is suitable equipment installed with PROINF/MDA funds for transporting animal production (live and slaughtered cargo), slaughtering and processing hides.

3 As an example, comparing the database provided by SEBRAE and the Agriculture Secretariats of the municipalities in the Médio Sertão region of Alagoas, the production of goats and sheep in the municipalities of the semi-arid region made available on the regional market, after the structuring of family production units, does not meet 1/5 of the per capita demand of family consumers looking for this type of product on the local market. As well as individual consumers, farmers can also count on the bar/restaurant trade and the town halls themselves to buy the products for inclusion in school meals.

Pigs and food supplements for 90 days.	02	
Purchase of poultry and complementary food for 90 days.	20	
String bean seeds (kg)	03	
Maize seeds (kg)	15	
Sorghum seed (kg)	03	
Zinc-plated seed storage tank with 120kg capacity	02	
Simplified irrigation kit and vegetable and fruit seedlings.	01	
Rustic sheepfold 40m2	01	
Appropriate fencing for the technologies (750 metres of barbed wire)	750	
Technical assistance (01 for 40 families)	01	
Work kit for ADLs (rucksack, boots, work uniform and tablet)	01	

* See technical description in ANNEX 1

Analysis

According to the logic of the project, it is understood that the productive character associated with the traditional farmyard is intended to provide fodder support for animals, by planting maize and sorghum, as well as contributing to the family's diet by planting cowpeas in conjunction with maize. In this sense, the water reserved in the reservoir will be used for irrigation[4] of maize, beans and sorghum crops. It's worth remembering that part of the area cultivated with both maize (13 %) and sorghum (25 %) will be "harvested" early for silage.

The project provided for the planting of 15kg of maize seeds, which was enough to plant 0.75 hectares. Part of the maize plantation was used to make silage mixed with sorghum and part was used for grain production, which was intercropped with cowpeas. This resulted in an approximate area of (0.45) ha cultivated exclusively with maize and (0.3) ha cultivated with maize and beans. Of the area planted with maize, 0.65 ha was used for grain production and (0.1) ha was used for silage production.

Therefore, for an availability of 2,340 kg of maize grain and 2,000 kg for silage with intercropping, 10 kg of cowpeas/ha were needed. As 3 kg of bean seed was planned, the area to be intercropped with maize was (0.3) ha, which produced 110 kg of beans. The 3 kg of sorghum seed was enough to plant 0.4 ha, of which (0.1) ha was for silage and (0.3) ha for grain production. The catchment area for the cistern was the sheepfold's 40 m roof^2 and it stores 24,000 litres of water, enough to water the animals during the

---

[4] Irrigation is carried out on rainfed crops when drought is anticipated or when "veranicos" occur,

dry season.

It can therefore be concluded from the above evidence that the water reserves provided by the component technologies of the system in question were sufficient to maintain the animals and minimise the risks in the production of fodder to feed them.

TABLE 4.1 - INTEGRATED PRODUCTION SYSTEM 02

INTEGRATED PRODUCTION UNIT	QUANT.	FAMILIES
Canvas yard with boardwalk with storage capacity (up to 108 thousand)	01	
Pigs and food supplements for 90 days.	03	
Purchase of shelter birds and food supplements for 90 days.	40	
String bean seeds (kg)	03	
Maize seeds (kg)	15	
Sorghum seed (kg)	03	**20 Families**
Seed storage tanks cap. 120 kg	02	
Simplified irrigation kit and vegetable and fruit seedlings.	01	
Appropriate fencing for the technologies (750 metres of barbed wire)	750	
Technical assistance (01 for 40 families)	01	
Work kit for ADLs (rucksack, boots, work uniform and tablet)	01	

* See technical description in ANNEX 1

Analysis

The difference between production systems 1 and 2 is the social water storage technology and its volume. The first system has a traditional reservoir for life-saving irrigation and the second has a tarpaulin reservoir. The project's traditional yard has the capacity to store up to 540m^3 of water, considering an average loss of 50% (evaporation and infiltration). In the case of the tarpaulin yard, it has a reserve capacity of 108 m^3 of water, i.e. less than !4 of the traditional yard. It should be borne in mind, however, that the losses due to infiltration and evaporation in the tarpaulin yard are minimal, considering that the reservoir is waterproofed with a HDPE membrane and will have a cover that should minimise evaporation losses.

In terms of predicting the volume of water stored, this system has less water security, but in the event of the need for life-saving irrigation, it is more secure in terms of preserving the initial volume over time than the tarpaulin.

In its complete cycle, maize consumes at least 400 mm/ha of water. Rainfall from March to June in the Ipanema region would be sufficient, but this rainfall is poorly distributed, and when a veranico occurs, it becomes necessary to irrigate to save the crop, providing just enough water to keep the plant alive. In agricultural activities, climate risk is a constant in any region, but in the semi-arid region it takes on greater proportions. It is not possible to predict or estimate the behaviour of rainfall in terms of volume and frequency during the cycle of the crops envisaged in the project. What is hoped is that the reserved water can reduce the uncertainties regarding rainfall and the effects of its lack on the production of fodder for animals.

In this context, it can be concluded that the system in question, considering the same simulations used to analyse the previous system, is capable of maintaining the animals and minimising the risks in producing fodder to feed them.

ADDITIONAL INFORMATION

Analysing the integrated production systems, it can be seen that sheep farming is the main component of production, which together with poultry farming makes it possible to generate monetary income for families. At the same time, the production of maize, beans and vegetables, as well as the fattening of pigs, should improve the families' food security. On the other hand, a municipal slaughterhouse for small animals is being built, which will make it easier to sell the animals when they are disposed of. This fact alone justifies the promotion of sheep farming, which ranks second in terms of animal numbers in the region, only behind cattle, which have a predominant position.

To support the introduction of sheep farming, the project also includes the training of technology disseminators for farmers, most of whom are new to this activity. According to the project, these ADLs (local development agents) will be recruited from high school students at local schools. It should be emphasised that these "trainees" will have to be very well qualified and trained in sheep and poultry management, according to the production systems envisaged in the project.

In addition to animal management, ADLs should be able to guide farmers in production planning and water management. When, where, how much and how to plant maize and sorghum, for grain and silage; likewise for beans, vegetable gardens and fruit trees. The less literate the recipients, the more necessary it is to make intensive use of technology to support communication, especially electronic audiovisual aids.

Although sheep farming is endemic to the region, it is estimated that the vast majority of beneficiary families are unfamiliar with this type of farming, which will require intense motivation and training work.

It was also necessary to obtain funds to cover the costs of basic animal health maintenance, especially vaccinations, dewormers and combating ectoparasites.

In order to speed up the process of spreading sheep farming, it would be advisable to set up a "demonstration unit" at a rural school or perhaps sign an agreement with an agricultural teaching institution in the region for continued training of family farmers in sheep farming.

ANNEX 1

A DESCRIPTIVE SYNOPSIS OF THE ANIMAL HUSBANDRY FACILITIES, RAINWATER HARVESTING EQUIPMENT AND FLOOR PLANS.

1) RUSTIC SHEEPFOLD

The APRISCO RÚSTICO is intended for sheltering and handling the goats and was installed by the families using existing materials on the property, such as round wood and carnauba straw for the roof, with a dirt floor. The materials purchased on the local market will be used to support the facilities, such as a masonry wall to support the slatted walls, feeders and a sanitising system.

The size of the sheepfold will be 40 metres^2 of usable area, structured in such a way as to allow the family producer to expand the facilities as the farm grows. The sheepfold has, essentially, four divisions for batches of animals in the following stages of development: goats in an advanced stage of pregnancy (close to giving birth) and newly calved goats; animals in the reproduction phase (matrices and breeding stock); goatherds (lactating animals) and weaned kids.

The first division should give access to a paddock with native or cultivated pasture. This area allows the goats close to giving birth and the newly calved goats to be properly managed, avoiding the action of predators and the occurrence of whipworms in the newborn animals.

Mineral salt troughs for supplementing the animals will be placed in each of the rooms reserved both for the batches of goats close to giving birth and those that have just given birth, as well as for breeding and weaned animals. The troughs can be made of tyres, planks or hollow logs found on the property and should be positioned at a height of 0.50 m from the ground. A guard, made of lath or wire, can be placed over them at a height of around 0.30 m above the height of the trough to prevent animals from entering.

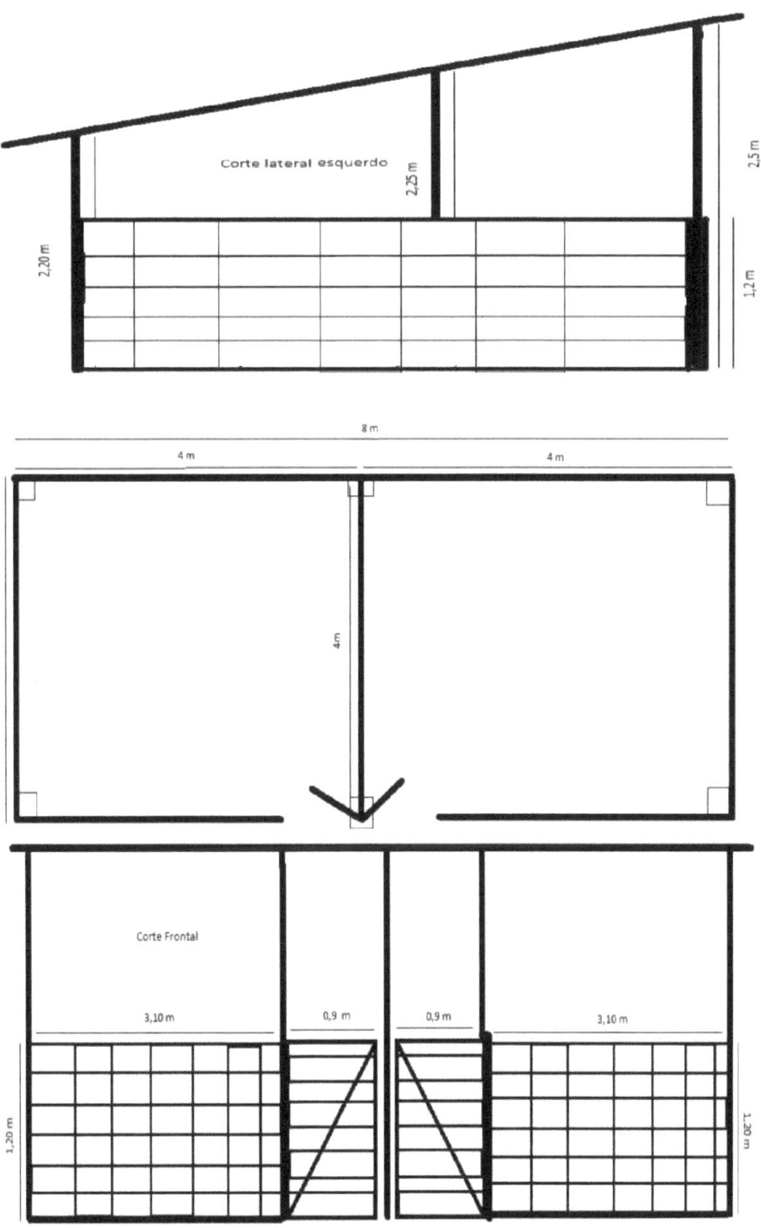

2) SHEEPFOLD CISTERN

Rural cisterns for collecting and storing rainwater in the semi-arid region of Alagoas are known as alternatives for meeting the needs of families during droughts. This social technology is also being used to store rainwater to meet the needs of the region's small herds. With a capacity of 24,000 litres, the cisterna aprisco (cistern for agricultural production) is installed next to the livestock farm and collects water from the roof of the sheepfold itself. The installation of the cistern accompanied by the water resource management course is an excellent socio-educational opportunity to make family farmers aware of the impact of overgrazing the caatinga that the extensive breeding system can have on the biome, increasing water scarcity and the risk of desertification. Workshops were held to train beneficiary families in the use of water, the importance of vegetables in their diet and livestock management, with a view to conserving the caatinga and water in the semi-arid region.

The methodology for building the sheepfold cisterns is similar to that used to install the slab cisterns for collecting water in family homes. The first step is to identify the most suitable area for installing the cisterns next to the sheepfold. Next, the area of the sheepfold's roof is measured to capture rainwater. Finally, the process of building the cistern begins, following these steps: marking the edge of the cistern, depth, excavation; demonstrating the templates and making the concrete slabs; making the floor; laying the slabs; tying the wall together; plastering; building and installing the top cover; preparing and installing the spouts; waterproofing, internal and external finishing, and fixing the identification plaques.

The cistern has the capacity to store 24,000 litres of water, enough to feed the animals provided for in the project. Part of the water from the cistern can also be used to irrigate the economic site, complementing the water support for production.

3) TARPAULINS WITH PAVEMENTS FOR SMALL FAMILY PRODUCTION

Canvas Cisterns are low-cost reservoirs for capturing and storing rainwater, which flows off a mobile boardwalk that will be built from the HDPE blanket itself, with a slope recommended for draining water from rural properties, as well as the destination of this water for the inside of the production cistern, The cisterns will have a roof and internal lining made of High Density Polyethylene (HDPE), a galvanised steel frame to support the roof, a filtering system made up of a fibre box and a polypropylene filter, and a water collection chute. In order for the rainwater to remain in good condition, the cistern must be made in such a way that it is well sealed, with no light entering, and if possible underground, thus reducing the temperature of the water and consequently slowing down the action of bacteria and the appearance of algae. Canvas Cisterns are manufactured in a single, whole, pre-made piece, ready to be applied by the farmer himself. The HDPE blanket is an excellent covering, providing insulation, flexibility, resistance to corrosion and abrasion, chemical resistance, mechanical strength, good weldability and a long service life. They will be excavated by machines and will have an asbestos tile roof and a catchment system using HDPE boardwalks. The tarpaulin yard is a social technology designed to capture and store rainwater. It consists of a trench-shaped reservoir excavated in the ground and waterproofed by applying HDPE geomembrane and covered with fibre cement tiles, with a waterproofed surface of the same material as the catchment area, with an area of 200 m^2 , complemented by rainwater run-off from its roof. The geomembrane for lining both the catchment area and the excavated reservoir is supplied by the manufacturer in single pieces, duly welded in the respective dimensions and formats, simplifying the application process.

It works by capturing rainwater in the area sealed with the HDPE geomembrane, which flows by gravity into a decanter to remove possible impurities, and is then channelled into the reservoir. The water captured by the roof of the reservoir is collected by a gutter and also channelled into the reservoir, which is capable of storing around 108 m^3 of water. The catchment area must be properly fenced off with sturdy mesh to prevent access by domestic animals, including dogs and cats, not only to ensure the cleanliness of the site, but also to protect the geomembrane from trampling, which could damage it.

It is used to collect and store rainwater and is of excellent quality in terms of turbidity, salinity and other physical and chemical characteristics. However, some microbiological contamination can occur, through insects, bird droppings, etc., which makes it impossible to use it for human consumption without treatment. If it is of good quality, it should primarily be used to water animals, but it can also be used for domestic purposes, except for drinking.

Procedures for installing the tarpaulin cistern:

- Defining the area for installing the cistern near the catchment;
- Excavate the ground, starting with the internal dimensions (length, width and depth) and then excavate the sloping walls, which will give the external dimensions (never

start by excavating the external dimensions);

- To fix the HDPE tarpaulin (anchor), it will be necessary to excavate a channel (trench) with a width of 30cm and a depth of 30cm. This channel will be dug at a distance of 40cm from the edge of the CISTERN excavation;
- Install the tarpaulin over the excavation;
- Once the tarpaulin has been installed and stretched, place the "leftover" blanket in the channels, it's time to fix the CISTERN in the channels. Return the removed soil to the channel, compacting it with a tamping tool;

- Lay the water inlet or outlet pipework at the bottom or side of the tank. Pipes should be exposed (appearing inside the tank) by no more than 20 cm (which is sufficient for sealing).
- Assembling the cover. Start assembly by connecting the metal parts of the roof as indicated, tightening the screws by hand;
- Once the structure has been assembled, the place where the access door to the CISTERN is to be built is defined.

- Installing the cover blanket. Place the blanket next to the CISTERNA structure;
- Once the covering blanket has been installed, the access door to the inside of the CISTERN is cut;
- Installation of the rainwater filtration system consists of: DISPOSAL BOX: eliminates the first rainwater, which will bring all the dirt that is on the roof; PRE-FILTER: serves to retain coarser sediments and works by the principle of decantation; FILTER: necessary to filter out the smaller particles still present in the water.

4) SMALL CLAY PITS

The construction of BARREIROS is an ancient practice in the semi-arid northeast of

Brazil for storing rainwater. This social technology is an alternative for farmers to accumulate water on their properties for use in small areas when drought threatens the loss of food crops and livestock. It is also mechanised in the flood shaft. The most important procedure is identifying the most suitable location, coating/finishing and handling. The main thing to look out for is the installation site, which needs to be on a sloping point of the property so that gravity irrigation is possible.

Producers in the semi-arid region can use the water stored in the reservoir to save a crop or to grow a second crop, which will give them additional income. The reservoir with a capacity of 500m^3 , occupies an area of approximately 200m^2 to store enough water to supplement four hectares in a period of drought.

São José da Tapera/AL, July 2018.

EDILSON RAMOS DE LIMA
Author of the Project

I want morebooks!

Buy your books fast and straightforward online - at one of world's fastest growing online book stores! Environmentally sound due to Print-on-Demand technologies.

Buy your books online at
www.morebooks.shop

Kaufen Sie Ihre Bücher schnell und unkompliziert online – auf einer der am schnellsten wachsenden Buchhandelsplattformen weltweit! Dank Print-On-Demand umwelt- und ressourcenschonend produziert.

Bücher schneller online kaufen
www.morebooks.shop

Printed by Books on Demand GmbH, Norderstedt / Germany